Candy Math
Book 1

by Vonae Tanner

Table of Contents

Chapter 1
Square Numbers

This is a piece of candy.
It has a length of 1.

This is a *line* of candies.
It has a length of 3.

This is a *square* of candies.
It has a length of 3 and a width of 3. It takes **9** pieces of candy to make this square.

9 is a square number.

8 is not a square number.

10 is not a square number.

9 is a square number.

$$3 + 3 + 3 = 9$$

$$3 \times 3 = 9$$

When the same number is multiplied many times, like multiplying 3 by itself a hundred times, writing it out can be difficult. To make this easier, mathematicians developed exponents, a shorter way to represent repeated multiplication.

An exponent is a small number found at the upper right corner of a larger number, called the base. An exponent tells you how many times to repeatedly multiply the base. For example,

$$3^2 \text{ means } 3 \times 3$$
$$3^3 \text{ means } 3 \times 3 \times 3$$
$$3^4 \text{ means } 3 \times 3 \times 3 \times 3$$
$$3^{100} \text{ means } 3 \times 3 \times 3 \times 3 \times 3 \times 3..... \times 3.$$

3^2 is read "three squared."

This square has a length of 4 and a width of 4. It takes 16 candies to make this square so 16 is a square number.

$$4 \times 4 = 16$$

$$4^2 = 16$$

This is the next bigger square, it's a 5 x 5 square of candies. How many pieces of candy does it take to make this square?

It takes 25 candies, so 25
is a square number.

$$5 \times 5 = 25$$

$$5^2 = 25$$

Let's explore how square numbers increase. How many *more* candies will it take to make a 3 x 3 square into a 4 x 4 square ?

It will take 7 more candies.

How many more candies will it take to make this 4 x 4 square into a 5 x 5 square?

It will take **9** more candies.

How many more candies will it take to make this 5x5 square into a 6x6 square?

It will take 11 more candies.

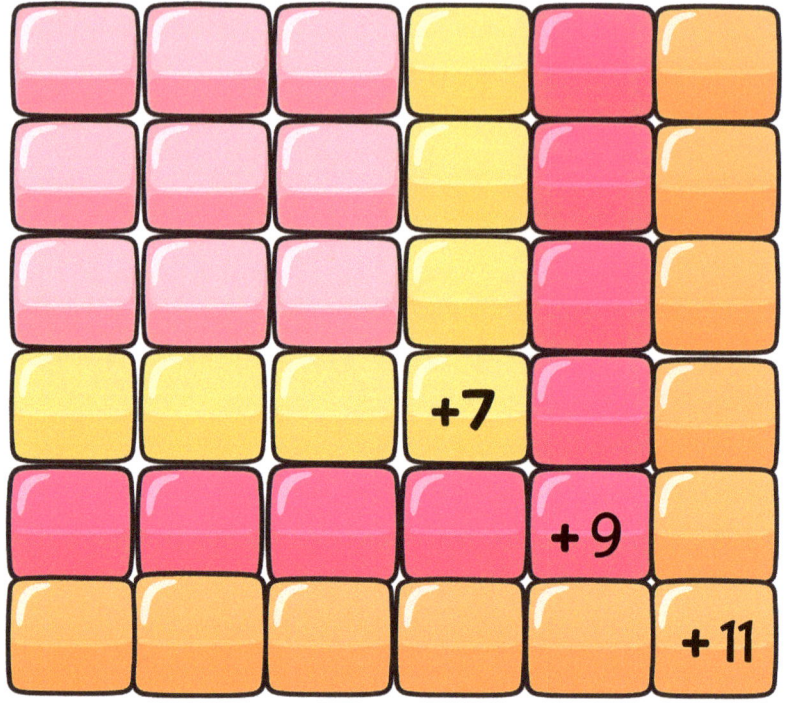

Why does it take an odd number of squares to make the next bigger square? Think about the corner piece.

0 1 4 9 16 25 ...

+1 +3 +5 +7 +9

The number you add to a square number to get the next bigger square number is called the difference. The pink numbers show the differences in the square numbers. The differences follow a pattern, they start at 1 and go up by 2 each time. This makes all of the differences odd numbers.

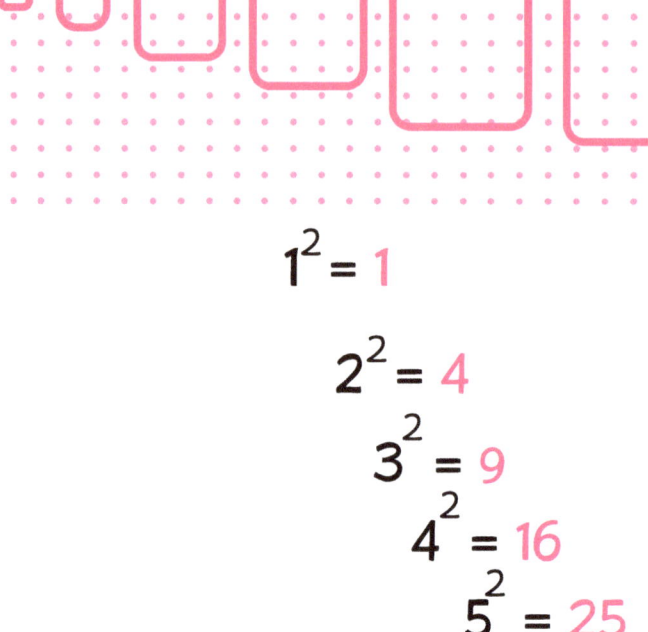

$$1^2 = 1$$
$$2^2 = 4$$
$$3^2 = 9$$
$$4^2 = 16$$
$$5^2 = 25$$

1, 4, **9** , 16, and 25 are square numbers.

Can you find more square numbers?

A number with an exponent of 2 is called "squared" because it can always create a perfect square. Likewise, every square can be represented by a number with an exponent of 2.

Challenge:

If a square is X candies long, how many candies will it take to make the square?

How many candies will it take to make the next bigger square?

Challenge:

If the length of a square is X, then it will take X^2 candies to make the full square.

$$(X)^2 = (X)(X)$$

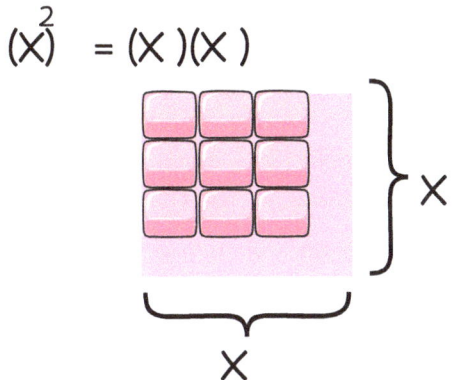

The length of the next bigger square will be (X + 1) and it will take $(X + 1)^2$ candies to make the full square.

$$(X + 1)^2 = (X + 1)(X + 1)$$

Challenge:

The number of candies needed to make a square with a length of (X + 1) is...

- the original square of X

+

x^2

$x^2 +$

- the original square of X

+

- 2 lengths of X

+

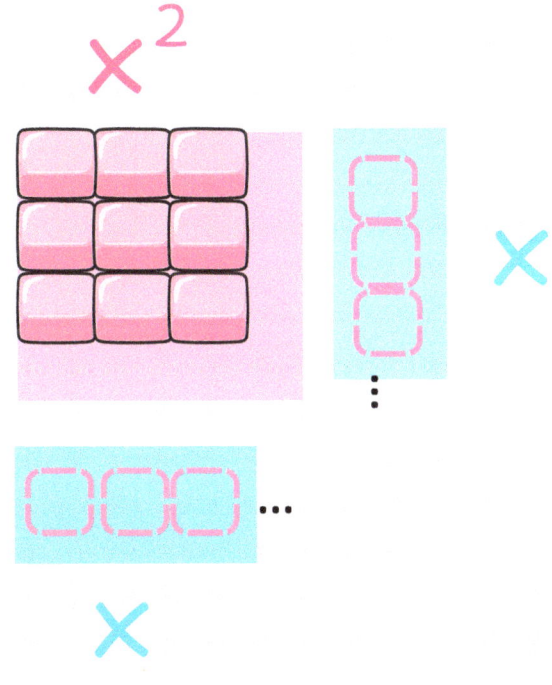

$$x^2 + 2X$$

Challenge:

- the original square of X

  +

- 2 lengths of X

  +

- 1 corner piece

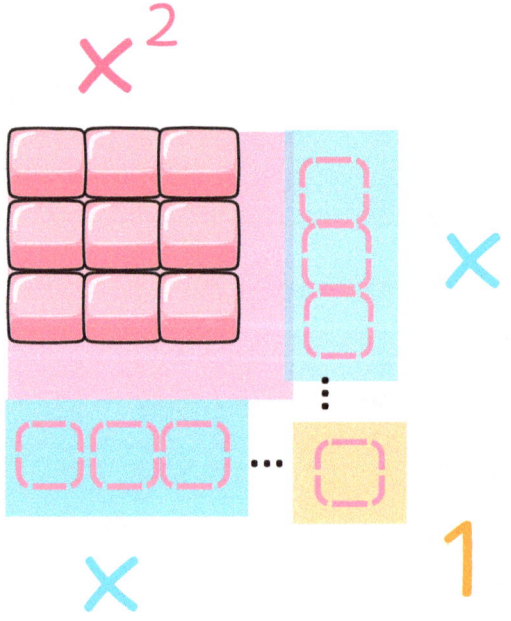

$$x^2 + 2x + 1$$

Challenge:

$$(X + 1)^2 = X^2 + 2X + 1$$
$$(X + 1)(X + 1) = X^2 + 2X + 1$$

History of Perfect Squares

Finding perfect squares is like a mathematical treasure hunt, one that has been going on for thousands of years. As early as 1600 BC, the ancient Egyptians were using square numbers. Around 1400 BC, Chinese mathematicians were recording them too, and by 800 BC, square numbers were appearing in Indian mathematics.

Galileo discovered a relationship between square numbers and falling things in his work on constant acceleration. He also discovered a relationship between odd numbers and falling things call the Law of Odd Numbers. The deep connection between these two discoveries can be found in the candy models you just made.

Even after all this time, square numbers still hold mysteries that mathematicians are working to solve. In 2019 , Philip Brown, a professor at Texas A&M University's Galveston Campus, created a new way to discover perfect squares. His method, or algorithm, is special because it only uses addition, subtraction, and multiplication, no division required! That's important because division can be tricky when working with large numbers. Even better, his algorithm can determine with 100% accuracy whether a number is a perfect square. This is something most other methods can't always guarantee.

There may still be secrets within square numbers and you might be the one to find them.

Chapter 2
Triangle Numbers

This is a piece of candy.
It has a length of 1.

This is a *line* of candies.
It has a length of 3.

This is a *triangle* of candies.
It has a base length of 3. It takes 6 pieces of candy to make this triangle.

6 is a triangle number.

3 + 2 + 1 = 6

This *triangle* has a base length o
4. It takes 10 pieces of candy to
make this triangle so 10 is a
triangle number.

$$4 + 3 + 2 + 1 = 10$$

How many more piece of candy are needed to make the next bigger triangle?

How many pieces will there be in all?

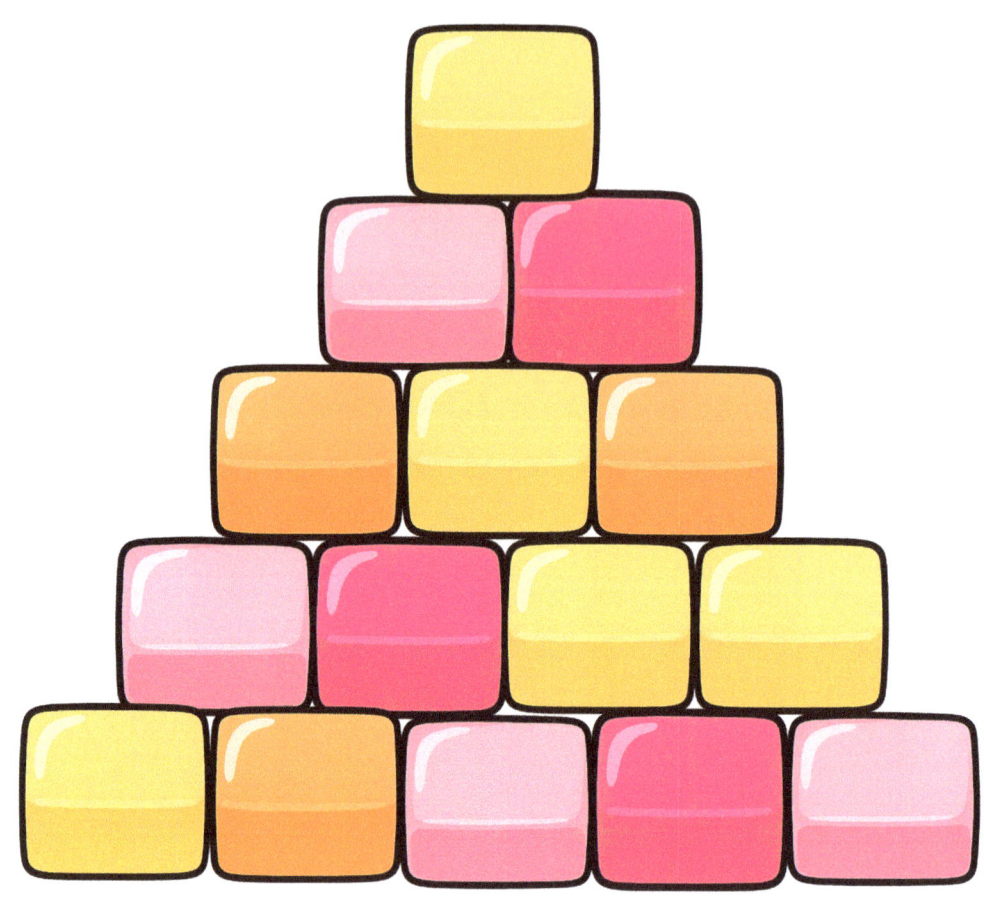

5 more pieces are needed to make the next bigger triangle, with 15 pieces in all, so 15 is a triangle number.

5 + 4 + 3 + 2 + 1 = 15

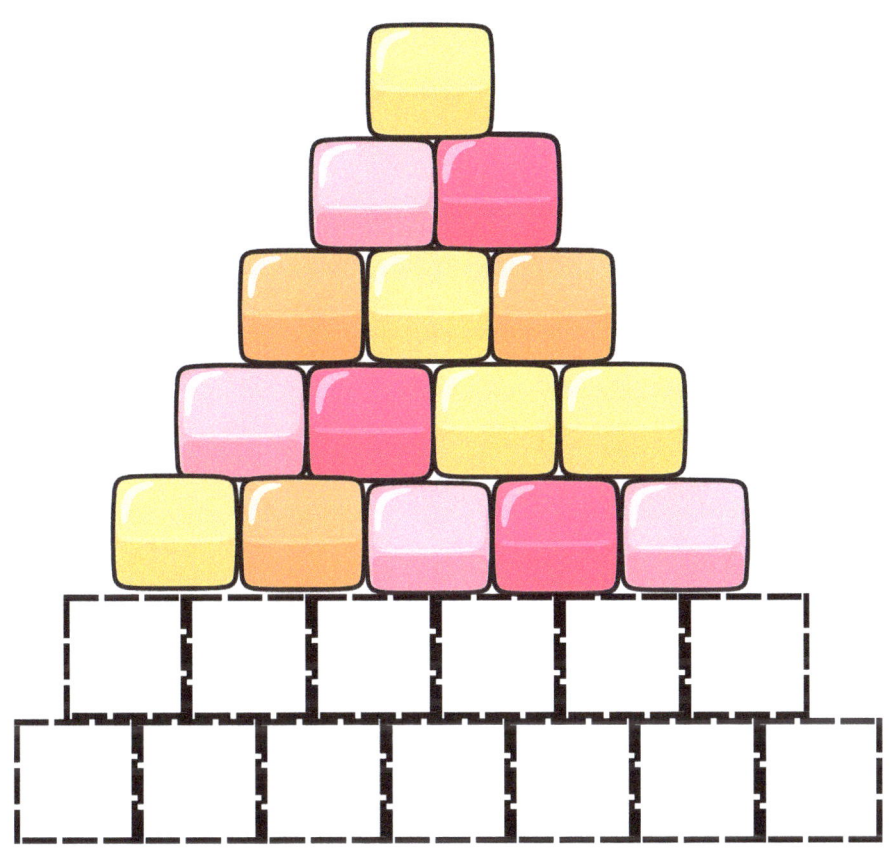

How many pieces of candy does it take to make a triangle with a base length of 6? What about 7?

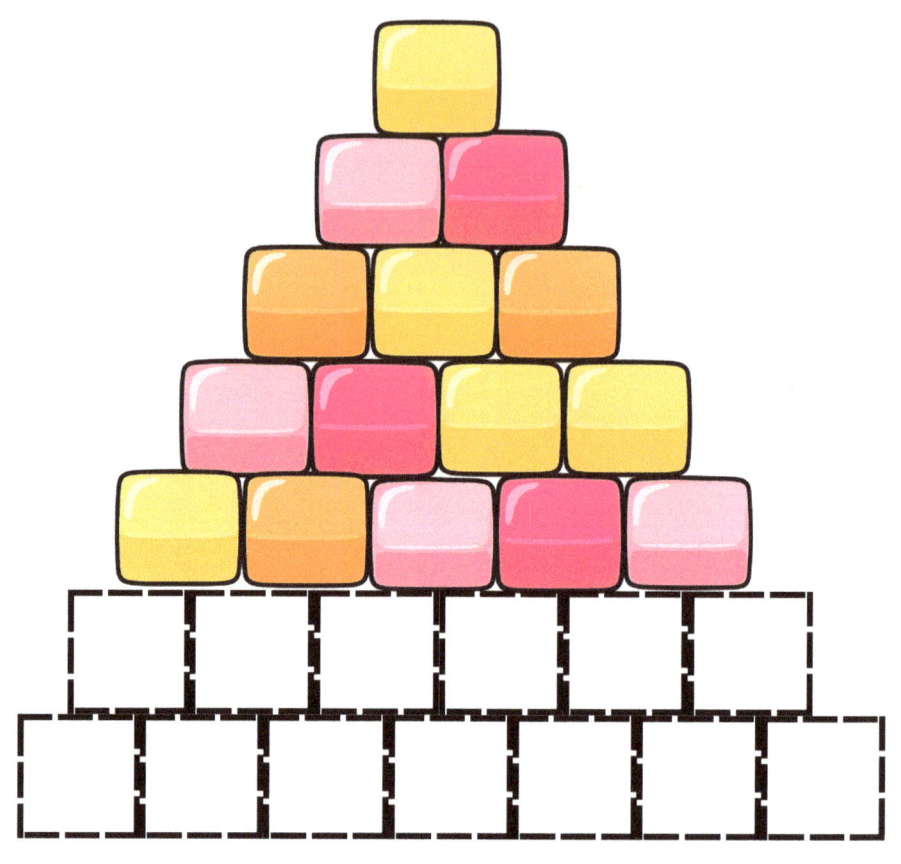

21 and 28 are triangle
numbers.

$$6 + 5 + 4 + 3 + 2 + 1 = 21$$
$$7 + 6 + 5 + 4 + 3 + 2 + 1 = 28$$

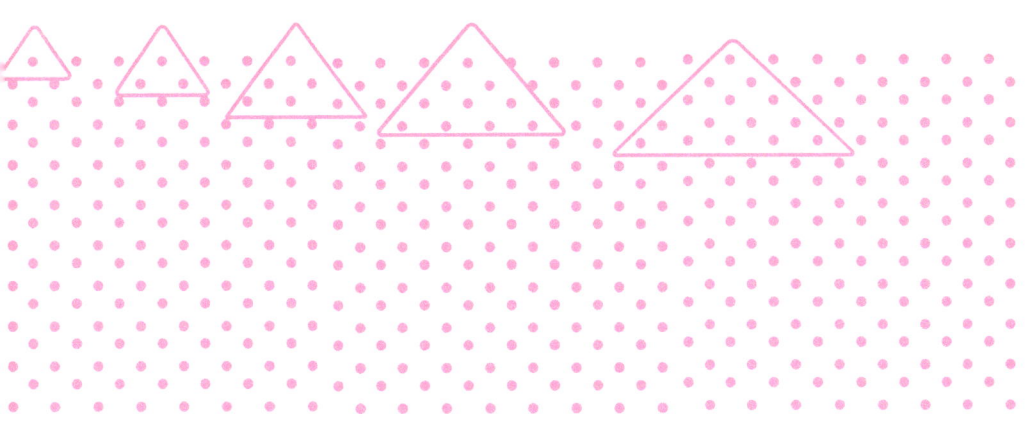

0, 1, 3, 6, 10, 15, 21, and 28 are triangle numbers.

Can you find more triangle numbers?

0   1   3   6   10   15   21 ...

+1 +2 +3 +4 +5   +6   +7

There is a pattern in the differences between each triangle number.

The differences increase by one.

Why? Because each new base is one candy longer than the previous base.

Challenge:

What do you get when you add any *two* consecutive triangle numbers (any two triangle numbers that come one after the other)?

How many numbers can be written as the sum of *three* triangle numbers?

Challenge:

The sum of any two consecutive triangle numbers is a square number!

$$3 + 6 = 9$$
$$6 + 10 = 16$$
$$10 + 15 = 25$$
$$15 + 21 = 36$$

All numbers can be written as the sum of three triangle numbers. Here are a few examples. Can you fill in the missing numbers? Can you find more?

$$0 + 1 + 1 = 2$$
$$0 + 0 + ? = 3$$
$$0 + 3 + ? = 4$$
$$3 + 1 + 1 = 5$$
$$6 + ? + ? = 19$$
$$21 + ? + ? = 39$$

The History of Triangle Numbers

The math symbol used to represent a sum of integers is the capital Greek letter sigma. It's used in summation notation as seen below. The summation notation below reads: let i represent the sum of integers from 1 to 4. This also happens to be the fourth triangle number, which is 10. Therefore the summation below equals 10.

$$\sum_{i=1}^{4} i = 1+2+3+4 \; = 10$$

Karl Gauss   discovered that every whole number can be represented as a sum of 3 triangle numbers. He also found an equation that can be used to find any triangle number. As the story goes, one day Gauss' teacher gave an assignment to add all the numbers from 1 to 100. After only a few minutes Gauss had the correct answer, stunning his teacher and classmates. How did he do it? He listed all the numbers from 1 to 100. Under those he listed all the number from 100 to 1. Then he simply added everything down and then up to get 10,100.

```
+  1   2   3   4   5   6   7   8   9  10  11  .... 97 98 99 100
  100 99 98 97 96 95 94 93 92 91 90 ..... 4  3  2  1
  101 +101+101+101+101+101+101+101+101+101+101  .... 101+101 +101 +101
```

Since he'd added from 1 to 100 twice he then divided his answer by 2 to get 5,050. Notice that adding 101 a hundred times is the same as multiply 100 by 101 so we could summarize his process by the following numeric expression: (100)(101) ÷ 2. This leads to the general formula for triangles numbers which is (n)(n+1) ÷ 2.

This equation can be used to find any triangle number. To find the 864th triangle number, plug 864 into the equation for n to get (864) (865) ÷ 2 which equals 373,680.

$$\sum_{i=1}^{n} i = \frac{n(n+1)}{2}$$

Chapter 3
Cube Numbers

This is a piece of candy.
It has a length of 1.

This is a *line of candies.
It has a length of 3.

A line is 1-dimensional.

(*A line of candy is not 1-dimensional but a line is.)

This is a *square* of candies. It has a length of 3 and a width of 3.

A square is 2-dimensional.

(*A square of *candy* is not 2-dimensional, but a square is.)

This is a cube of candy.
It has a length of 3, a width of 3, and a height of 3. It takes 27 piece of candy to make this cube.

A cube is 3-dimensional.

27 is a cube number.

$$3 \times 3 \times 3 = 27$$

$$3^3 = 27$$

3^3 is read "three cubed"

This is a cube. It has a length of 4, a width of 4, and a height of 4. It takes 64 candies to make this cube, so 64 is a cube number.

$$4 \times 4 \times 4 = 64$$

$$4^3 = 64$$

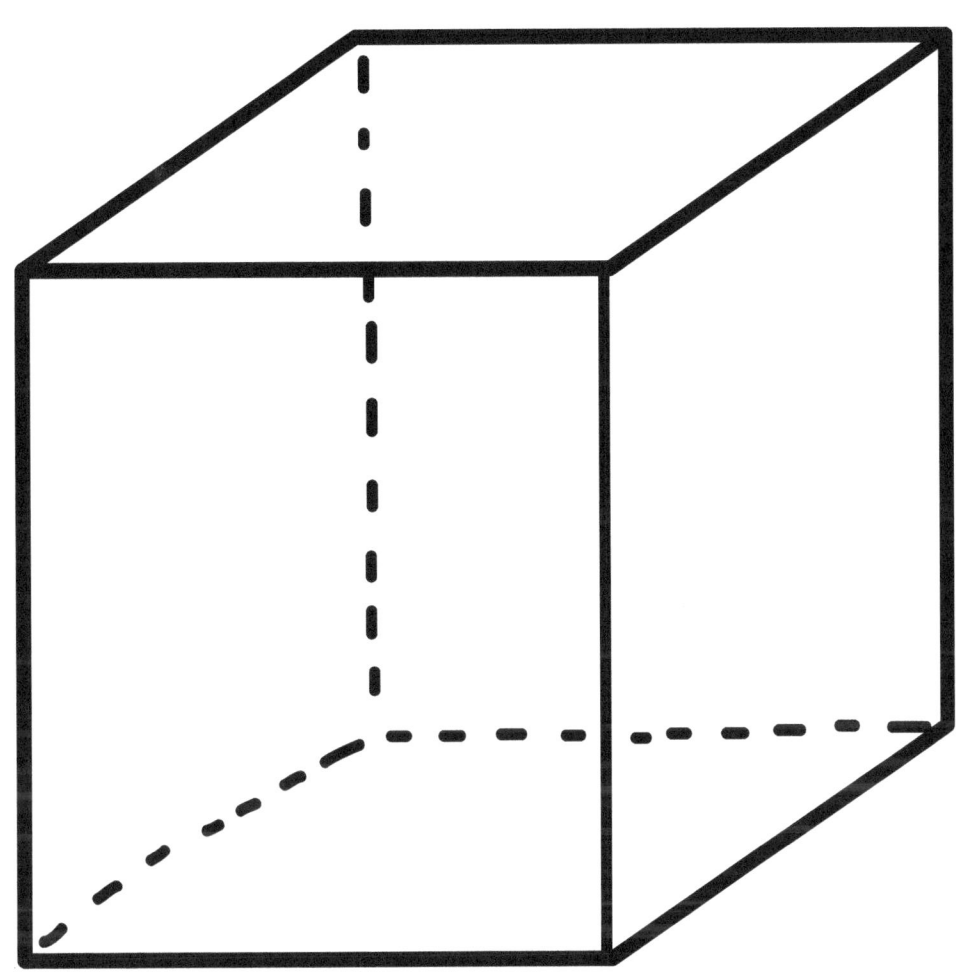

How many candies will it take
to build a 5 x 5 x 5 cube?

What is 5 cubed ?

0, 1, 8, 27, 64, 125 are all cube numbers. Can you find more cube numbers?

$$0^3 = 0$$

$$1^3 = 1$$

$$2^3 = 8$$

$$3^3 = 27$$

$$4^3 = 64$$

$$5^3 = 125$$

$$\vdots$$

$$0 \quad 1 \quad 8 \quad 27 \quad 64 \quad 125$$

$$+1 \quad +7 \quad +19 \quad +37 \quad +61$$

$$+6 \quad +12 \quad +18 \quad +24$$

The pink numbers are the differences between the cube numbers. They don't appear to follow a pattern. The orange numbers are the differences between the differences. They are called second differences. The second differences do follow a pattern.

Find and use the pattern to verify that the next three cube numbers are 216, 343, and 512.

Challenge:

How many candies would it take to build a 4-dimensional cube that has a length of 3?

Try to build it!

A 4-D cube with a length of 3 would require 3 x 3 x 3 x 3 candies, or 3^4 candies which is 81 candies.

It would have...
a length of 3
a width of 3
a height of 3
and a *what* of 3?

A 4-D cube is easy to **math**, but impossible to **build** in our 3-D world.

The History of Cubed Numbers

The Babylonians used perfect cube numbers to calculate volume as early as 1700 BCE. Around 500 BCE, Euclid wrote *Elements*, one of the first math textbooks, where he defined cube numbers and explored their properties. Later, around 60 AD, the Greek philosopher Nicomachus of Gerasa recorded an amazing pattern that connects cube, triangle, and square numbers. He found that the sum of the first n cube numbers equals the square of the nth triangle number. For example, the sum of the first four cube numbers is 36, and the fourth triangular number (6) squared is also 36. This pattern holds true for any value of n.

In math we can easily imagine a 1-D line, a 2-D square, and a 3-D cube, but visualizing a 4-D hypercube is almost impossible in three-dimensional space. Although 4-D *math* is straightforward, picturing 4-D shapes is tricky. Luckily there are many online animations showing how a 4-D hypercube would appear rotating through 3-D space. There are videos of 4-D hyperspheres and other 4-D shapes and even 11-D shapes such as the Calabi-Yau Manifold.

Numbers exist that represent other 3-D shapes, like the tetrahedral, or pyramid numbers. To build pyramid numbers stack consecutive squares on top of each other, with the top square being 1. Building a model of pyramid numbers may reveal why the sum of the first n cubes equals the square of the nth triangular number. See if you can discover the connection!

Chapter 4
Fibonacci Numbers

This is a piece of candy.
It is a square.

Adding 1 more piece of candy makes a rectangle.

Adding 2^2 more candies
makes a larger rectangle.

Adding 3^2 more candies makes a larger rectangle.

To make the next bigger
rectangle, 4^2 more
candies is not enough,
5^2 more candies are
needed.

To make the next bigger rectangle, 8^2 more candies must be added. How many are needed to make the next bigger rectangle?

Adding 13^2 more candies will make the next larger rectangle.

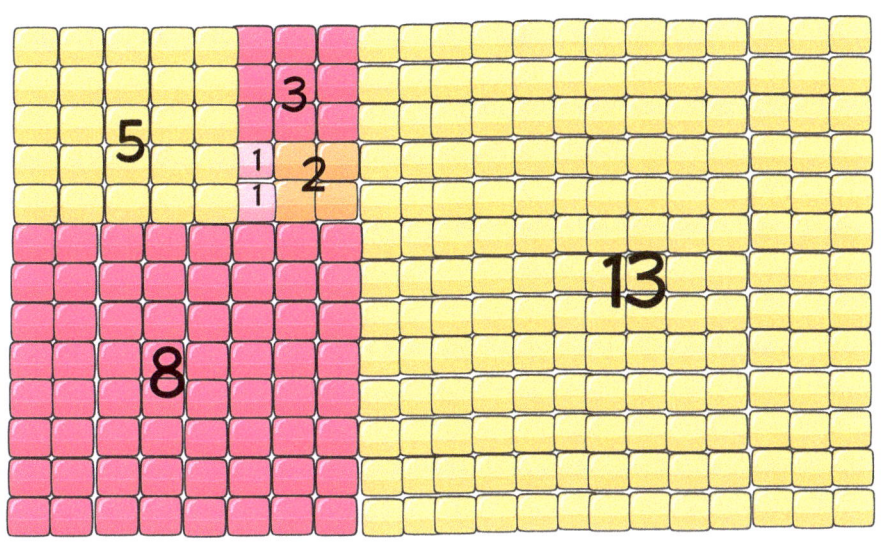

The side lengths of each square inside this rectangle are 1, 1, 2, 3, 5, 8, 13...

These are called Fibonacci numbers.

Fibonacci Numbers: 1, 1, 2, 3, 5, 8, 13...

$$1 + 1 = 2$$
$$1 + 2 = 3$$
$$2 + 3 = 5$$
$$3 + 5 = 8$$
$$5 + 8 = 13$$
$$\vdots$$

The sum of two consecutive Fibonacci numbers is the next Fibonacci number.

Find the next three Fibonacci numbers.

Fibonacci numbers show up in surprising places, like pinecones!

If you look closely at a pinecone, you'll see spiral patterns going in clockwise and counterclockwise directions. Surprisingly, the number of spirals in each direction is always a Fibonacci number. This pinecone has 8 clockwise spirals and 13 counterclockwise spirals!

Fibonacci numbers are found in the spirals of pineapples, sunflowers, succulents, and even broccoli.

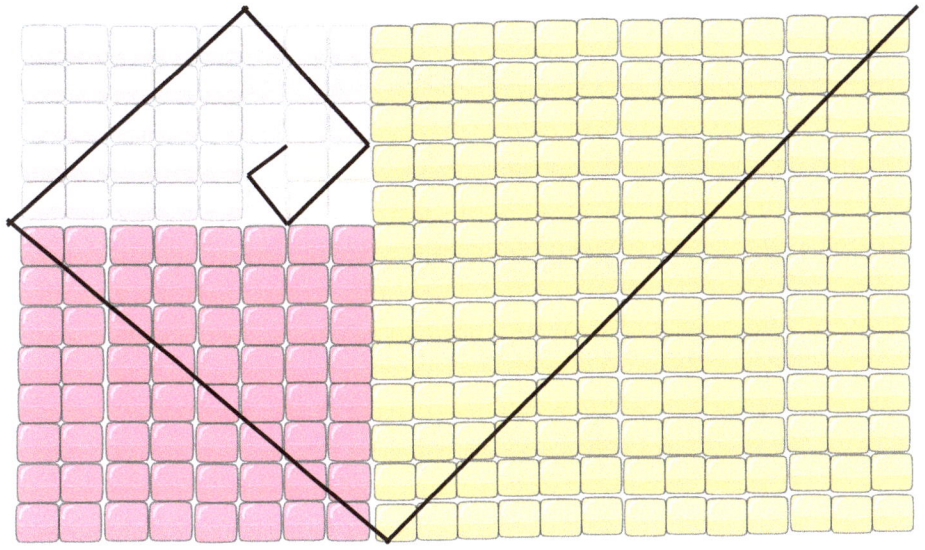

This rectangle is an approximation of a Golden Rectangle, which is found in famous art and architecture. Drawing a diagonal line through each Fibonacci square reveals a spiral pattern called the Fibonacci Spiral.

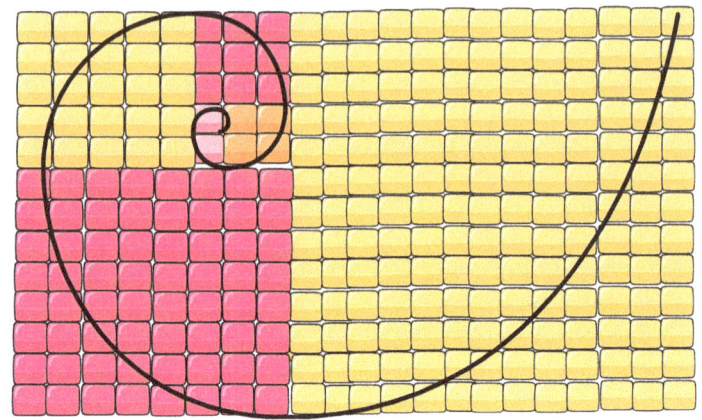

Rounding the diagonal lines creates a smooth spiral known as the Golden Spiral.

Golden Spirals provide the structure in seashells, hurricanes, sunflowers, snails, broccoli, and even sneezeworts.

Golden Spirals

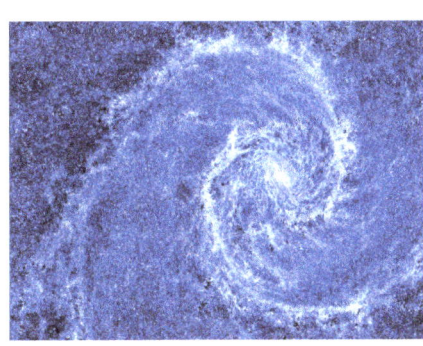

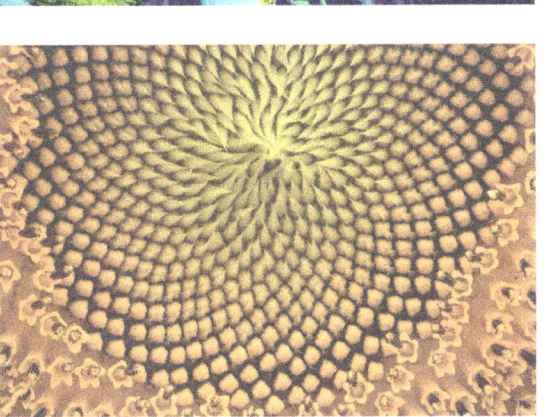

Challenge:

Find the ratios of Fibonacci Numbers by dividing a Fibonacci number by the one before it.

What pattern do you see?

Challenge:

1/1 = 1

2/1 = 2

3/2 = 1.5

5/3 = 1.666

8/5 = 1.6

13/8 = 1.625

21/13 =1.61538

When dividing consecutive Fibonacci Numbers an interesting pattern appears. The ratios bounce back and forth, getting closer and closer to 1.618... a famous irrational number called Phi, also known as the Golden Ratio.

The first few digits of Phi are

1.6180339887498948482045...

The History of Fibonacci Numbers

Fibonacci numbers are named after a famous mathematician who lived during the Middle Ages, Leonardo Fibonacci of Pisano. He posed a problem about rabbits and the solution led to this number pattern that now bears his name.

The number Phi was known *before* the Fibonacci numbers. During the Renaissance, artists like Leonardo da Vinci used Phi in their work, believing it created perfect beauty and balance. The connection between Fibonacci numbers and Phi wasn't fully understood until much later.

Artists and architects have used the Golden Ratio for centuries to create beautiful and balanced designs. You can find Golden Rectangles in famous paintings like the Mona Lisa and the Vitruvian Man, in architecture such as the Parthenon, in music by Bartok, in the human face, and in the rotational angles of leaves on a stem.

One final note on the golden ratio Phi, it has a fascinating mathematical property. It can be expressed by the infinitely continuous fraction seen below.

$$\varphi = 1 + \cfrac{1}{1 + \cfrac{1}{1 + \cfrac{1}{1 + \cfrac{1}{1 + \ddots}}}}$$

Chapter 5
Fractals

This is a piece of candy.
It has a length of 1.

This is a line of candies.
It has a length of 3.

This is a square of candies with the center piece removed. It's an *open* square of candy. It has a length of 3 and is made of 8 pieces of candy.

This is an open square of 8 open squares of candy. It has a length of 3 open squares. It is made of 64 pieces of candy.

This is an open square made of 8 open squares, which are made of 8 open squares of candy. It has a total of 512 pieces of candy. This is a fractal.

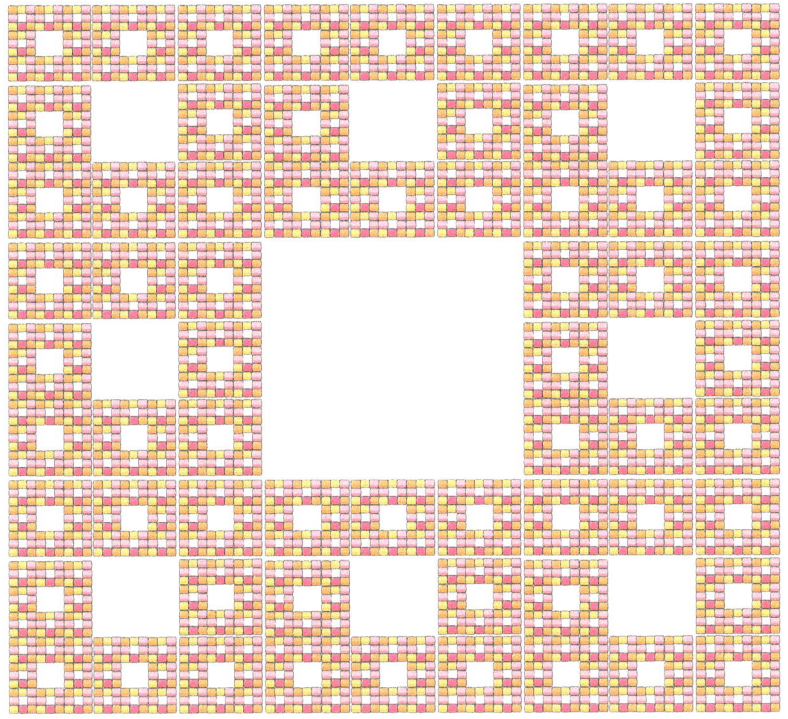

This fractal is called the Sierpinski Carpet. Fractals are self-similar, meaning they appears the same at every scale. Whether you zoom in on a single open square or zoom out to view the whole pattern, the structure remains similar.

Challenge:

1, 8, 64, 512, are the number of candies needed to make the first four iterations (or repetitions) of the Sierpinski Carpet. Can you write them using exponents?

Can you find the next three Sierpinski Carpet numbers?

The numbers in the Sierpinski Carpet pattern can be expressed using exponents with a base of 8.

$$8^0 = 1$$
$$8^1 = 8$$
$$8^2 = 64$$
$$8^3 = 512$$

The next 3 Sierpinski Carpet numbers are:

$$8^4 = 4{,}096$$
$$8^5 = 32{,}768$$
$$8^6 = 262{,}144.$$

The History of Fractal Geometry

Fractal geometry started when Edward Lorenz was studying weather patterns in the 1960s. He discovered that tiny changes in a weather simulator led to huge differences later on, an idea now called the butterfly effect. From this finding the field of Chaos was born. Chaos looks at how complicated systems behave in unpredictable ways. As scientists explored chaos, they noticed repeating geometric patterns called fractals. One early example is the Koch snowflake, made by adding smaller triangles to each side of a larger triangle over and over, creating a shape with an endless perimeter but a limited area. This makes the dimension of the Koch Snowflake about 1.26186.

Mathematically, fractals are fascinating because they exist in a decimal dimension and they use patterns that repeat forever. One famous fractal was created by Benoit Mandelbrot and is based on an equation involving complex numbers. The following are a few images from the Mandelbrot fractal.

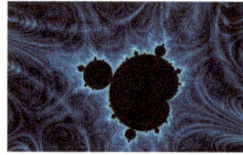 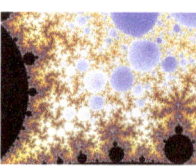 

The candy fractal in this chapter is called the Sierpinski Carpet, named after Waclaw Franciszek Sierpinski, a mathematician born in Poland in 1882. Like other fractals, it can go on forever. The Sierpinski carpet also has a decimal dimension, isn't 1-dimensional or 2-dimensional but 1.8-dimensional.

There is a 3-D version of the Sierpinski carpet as well. It's called the Sierpinski Sponge, or the Menger Sponge. If you have enough candy, try to make one! Build a 3x3x3 cube out of candy, and remove the center piece from each face and from the very center of the cube. This is the building block of a Sierpinski Sponge. Make 20 of these and put them together as a cube of cubes, with the center cube removed. With enough candy you could iterate (repeat) this process again and again, eventually making a Sierpinski Sponge big enough to walk through!

Fractals are used today in many fields. From cell phone antennas to graphic design to medical imaging and financial analysis, fractals model real-world patterns that at first glance seem random and chaotic but are actually governed by beautiful patterns.

In this journey through Candy Math we've built square numbers, triangle numbers, cube numbers, Fibonacci numbers, and fractal numbers. Along the way, we've uncovered deeper treasures like patterns hidden in sequences, the mystery of four-dimensional hypercubes, and the elegance of self-similarity. We've discovered the magic of Fibonacci numbers, the perfect proportions of the Golden Rectangle, the graceful curves of Golden spirals, and the incredible power of exponential notation. But this is just the beginning.

Math is alive everywhere you look. Keep looking, keep imagining, keep creating. Every wonder you explore and each model you create brings you closer to understanding the incredible, mathematical world we live in.

"Imagination is more important than intelligence." - Einstein

Check out more Candy Math adventures at www.mathmindworkshop.com! There are endless wonders yet to be uncovered.

www.ingramcontent.com/pod-product-compliance
Lightning Source LLC
Chambersburg PA
CBHW051232120626
46547CB00013B/1605